Studies in Logic
History of Logic
Volume 121

The Summa Ekliptika

The Cognitive Dimension of Social Argumentation. Proceedings of the 4th European Conference on Argumentation, Volume I. Fabio Paglieri, Alessandro Ansani and Marco Marini, eds.

Volume 110
The Cognitive Dimension of Social Argumentation. Proceedings of the 4th European Conference on Argumentation, Volume II. Fabio Paglieri, Alessandro Ansani and Marco Marini, eds.

Volume 111
The Cognitive Dimension of Social Argumentation. Proceedings of the 4th European Conference on Argumentation, Volume III. Fabio Paglieri, Alessandro Ansani and Marco Marini, eds.

Volume 112
Implicative-groups vs. Groups and Generalizations. Second Edition
Afrodita Iorgulescu

Volume 113
Meaning as a Set-theoretic Object. A Gentle Introduction to the Ideas Behind Formal Semantics
Jaroslav Peregrin

Volume 114
Logic and Language: New Frontiers in Analysis and Interpretation
José David García-Cruz, ed.

Volume 115
Four Functors. The 'ancestor' of Mundici's equivalence between unital commutative *l*-groups and MV algebras
Afrodita Iorgulescu

Volume 116
Logic: A Primer
Neil Tennant

Volume 117
Computational Paths
Arthur Freitas Ramos, Ruy J.G.B. de Queiroz, Anjolina Grisi de Oliveira and Dov M. Gabbay

Volumes 118-120
Argumentation in the Digital Society. Proceedings of the 5th European Conference on Argumentation, Volumes I, II and III
Jakub Pruś, Marcin Koszowy and Miriam Kobierski, eds.

Volume 121
The Summa Ekliptika
Manuel González Riqulme

Studies in Logic Series Editor
Dov Gabbay dov.gabbay@kcl.ac.uk

The Summa Ekliptika

Manuel González Riquelme

ISBN 978-1-84890-518-4

College Publications
Scientific Director: Dov Gabbay
Managing Director: Jane Spurr

http://www.collegepublications.co.uk

Cover prepared by Laraine Welch

CONTENTS

PREFACE

The insect “stick” is a phasmid that is mimicked with the plants due to his branch aspect.

Human beings are incarnations of theories about of laws of nature extracted of the portion of nature with the one that they were:

El tamaño y la fortaleza de las alas de un ave reflejan la fuerza intrínseca de la gravedad, sobre la cual el ave no tiene conocimiento teórico. La estructura de nuestros ojos y oídos encarnan verdades acerca de los fenómenos que llamamos “luz” y “sonido”, independientemente de nuestras teorías y creencias sobre ellos[1].

Nature obeys rules (or patterns) and it is also subject to limit: There are things that, evidently, cannot happen. The natural rules allow us to build rules of complexity every bigger time. This is a consequence of “life” that interacts with other rules. It is given an inevitability of norm for any cognoscible universe which implies that the descriptions of these rules can exist. Even as the classes of Russell, rules can exist in a group of rules and rules in the rules and, this way successively:

Para describir estas pautas necesitamos un catálogo de todas las pautas posibles y a ese catálogo lo llamamos *matemática*[2].

I add the Logic. Phasmid Logic.

---

1 John D. Barrow (1994: 283).
2 John D. Barrow (1994: 283).

I

Limited and Unlimited Indeterminism

Borges is who expresses better in their poem *The Causes* the principle of causation:

The sunsets and generations.
The days and none was the first.
The coolness of water in Adam's
throat. Orderly Paradise.
The eye deciphering the dark.
The love of wolves at dawn.
The word. The hexameter. The mirror.
The Tower of Babel and pride.
The moon that Chaldeans gazed at.
The innumerable sands of the Ganges.
Chuang-Tzu and the butterfly that dreams him[1].
The golden apples on the islands.
The steps in the wandering labyrinth.
Penelope's infinite tapestry.
The Stoics' circular time.
The coin in the dead man's mouth.
The weight of the sword on the scale.
Each drop of water in the clepsydra.
The eagles, the auspicious days, the legions.
Caesar on the morning of Pharsalia.
The shadow of the crosses over the earth.
The chess and algebra of the Persian.
The footprints of long migrations.
The conquest of kingdoms by the sword.
The relentless compass. The open sea.
The clock's echo in memory.
The king beheaded by the ax.
The incalculable dust which was armies.
The nightingale's voice in Denmark.
The calligrapher's meticulous line.
The face of the suicidal one in the mirror.
The gambler's card. Greedy gold.
The shapes of a cloud in the desert.

---

[1] Chuang Tzu dreams that he was a butterfly and he didn't know to the awakening if he was a man that dreamt a butterfly or a butterfly that dreamt a man.

Every arabesque in the kaleidoscope.
Each regret and each tear.
All those things were necessary
so that our hands would meet[2].

Leibniz Writes that there are two famous labyrinths in which our reason is often lost.

One of them is the problem of freedom and necessity, and the other is concerned with continuity and infinity[3].
Prior mentions to Jonathan Edwards:

What is self-existent must be from eternity...: but as to all things that *begin to be,* they are not self-existent and therefore must have some foundation of their existence without themselves. That whatsoever begins to be, which before was not, must have a Cause why it then begins to exist, seems to be the first dictate of the common and natural sense which God hath implanted in the minds of all mankind[4].

And later,

It is... as repugnant to reason, to suppose that an act of the will should come to existence without cause, as to suppose the human soul, or an angel, or the globe of the earth, or the whole universe, should come into existence without a cause. And if once we allow that such sort of effect as a Volition may come to pass without a cause, how do we know but many other sorts of effects may do so too?[5]

The determinist looks:

The determinist looks at the events taking place in the world as if they were a film drama produced in some cinematographic studio in the universe. We are in the middle of the performance and does not know its ending, although each of us is not only a spectator but also an actor in the drama. But the ending is there, it exists from the beginning of the performance for the whole picture is completed from eternity. In it all our parts, all our adventures and vicissitudes of life, all our decisions and decide, both good and bad, are fixed in advance. Even the moment of our death, of yours and mine, is laid down beforehand. We are only puppets in the universal drama. There remains for us nothing else to do but watch the spectacle and patiently await its end.

---

[2] Jorge Luis Borges, *Historia de la noche* (1977: 165, 166).
[3] Jan Łukasiewicz, "On Determinism" (1961: 30).
[4] Arthur Norman Prior, *Papers on Time and Tense* *(PTT)* "Limited Indeterminism" (2003: 73).
[5] Arthur Norman Prior, *PTT* (2003: 76).

This is a strange view and by no means obvious. However, there are two arguments of considerable persuasive power which have been known for a long time, and which seem to support determinism. One of them, originating with Aristotle, is based on the logical principle of the excluded middle, and the other, which was known to the Stoics, on the physical principle of the causality[6].

There is an error in the argument which derives the thesis of determinism from the principle of causality:

Let us consider time as a straight line and let us establish a one-to-one correspondence between certain interval of time and the segment (0,1) of that line. Let us assume that the present instant corresponds to point 0, that a certain future fact occurs at instant (corresponding to point) 1, and that the causes of this fact occur at instants determined by real numbers greater than ½. This sequence of causes is infinite and has no beginning, that is, no first cause. For this first cause would have to take place at the instant corresponding to the smallest real number greater than ½, and no such that number exists; not even the smallest rational number greater than ½ exists. In the set of real numbers and similarly in the ordered set of rational numbers, there are no two numbers succeeding each other immediately that is, being immediate predecessor and successor of each other; between any two numbers there is always another one, and consequently there are infinitely many numbers between any two of them. Similarly, there are no two instants succeeding each other immediately, that is, being the immediate predecessor and successor of each other; between any two instants there is another one, and consequently there are infinitely many instants between any two of them. In accordance with the principle of causality, every fact of the sequence under consideration has its cause in some earlier fact. Although it has a lower limit at instant ½, which is later than the present instant 0 and has not yet been attained, the sequence is infinite. Furthermore, the sequence cannot exceed its lower limit and therefore cannot reach back to the present instant.

This reasoning shows that there might exist infinite causal sequences which have not yet begun, and which belong entirely to the future. This view is not only logically possible but also seems to be more prudent than the belief that each, even the smallest, future event has its causes acting from the beginning of the universe. Therefore, the argument based on the principle of causality falls to the ground. One can be strongly convinced that nothing happens without cause, and that every fact has its cause in some earlier fact, without being a determinist[7].

Neither is the Principle of the Excluded Middle:

To me, personally the principle of bivalence does not appear to be self-evident. Therefore, I am entitled not to recognize it and to accept the view

---

[6] Jan Łukasiewicz, "On Determinism" (1961: 22, 23); see also Peter Øhrstrøm y Per Hasle (1995: 149-154).

[7] Jan Łukasiewicz, "On Determinism" (1961: 30-32).

that besides truth and falsehood there exist other truth-values, including at least one more, the third truth value.

What is this third value? *Indeterminate.* All sentences about future facts which are not yet decided to belong to this category. Such sentences are neither true at the present moment, for they have no real correlate, nor are the false, for their denials too have no real correlate. We could say that ontologically there corresponds to these sentences neither being nor non-being but possibility. Indeterminate sentences which ontologically have possibility as their correlate, take the third truth-value.
This trivalent logic is as consistent and free from contradictions as is bivalent logic. Whatever form, when worked out in detail, this new logic assumes, the thesis of determinism will be no part of it[8].

Prior point out [in response to Jonathan Edwards] that is the nature of "volitions", which makes their coming-to-be-possible but nor necessary [causes, events], and the nature of their non-existent or not-yet-existent occurrences which makes their coming-to be either necessary or impossible.

In the strict sense of "are", there "are" no actions and no capacities but things that act and things that are capable of acting. I repeat, it is *their* limitations -the limitations of actually present things, not those of still absent events- which, while leaving some alternatives possible, do not leave all alternatives possible.

We have done it by falling back on a metaphysic of "substances" endowed with capacities and dispositions, and our line of escape would hardly be open to someone (let us say Professor Hartshorne) who believes that objects are logical constructions out of events rather than vice versa. And it seems to me that this apparent dependence of the distinction between a limited and unlimited indeterminism on this metaphysic of substances endowed with capacities, is a definite argument in favor of such metaphysic. For the distinction between limited and unlimited indeterminism *is* one which it seems prima facie possible to make[9].

---

[8] Jan Łukasiewicz. "On Determinism" (1961: 36, 37).
[9] Arthur Norman Prior, *PTT* (2003: 78).

II

Anisotropic Time

For Horwich:

Thus, our position is like that of ornithologist who has looked thoroughly, but in vain, for a White raven and infers that there aren´t any. Similarly, it seems fair to conclude, from what we know now, that time is probably isotropic[10]

But the fact that now with progressively later events indicate nothing about structure of time:

It does not imply that time has an "arrow" or a directional character of any sort. For, as we have seen, "now" is an indexical expression, line "here", so there is no metaphysical significance in the variable location of its referent. However, there remains the possibility that time is intrinsically asymmetric. It still could be that time exemplifies some asymmetry, having little to do with the behavior of *now* but somehow related to patterns of events that are asymmetric with respect to time. Perhaps, in other words, the glaring time biases displayed by phenomena such as knowledge, causation, and dissipation of order point the way toward an asymmetry within the structure of time itself. It could be that the two opposite directions along the temporal dimension are significantly different from one another, even though neither has the metaphysically special status of being the direction "in which time goes". Thus, we should open the idea that time is anisotropic, despite having no *privileged* direction[11].

Despite that there are temporal asymmetric events. Horwich differences, at least ten:

(1) Now: We have a sense that time flows from past to future. (2) Truth: Following Aristotle, it is often maintained that contingent statements about future have no truth value, unlike statements concerning the past and present which are determinately either true or false. (3) Laws: Suppose there is a process whose temporal mirror image is impossible –that is, a sequence of states of affairs, *ABCD,* where the reverse sequences, *DCBA*, would be ruled out by laws of nature. In such cases *ABCD* is said to be time-nomologically irreversible, and the operative laws are said to be time-asymmetric. Consider the second law of thermodynamics, which states that the entropy of an isolated system will never decrease whose

[10] Paul Horwich (1992: 57).

[11] Paul Horwich (1992: 37).

temporal inverse are ruled out by it. (4) De facto, irreversibility: Schematically the sequence of states *ABCD* is common, whereas instances of *DCBA* are extremely rare. The central problem presented by these de facto one-way processes is to *explain* the temporal asymmetry that they display. It will be seen that laws of nature are not enough to account for it. (5) Knowledge: we know more about the past than we know about future. (6) Causation: Effects never seem to precede their causes. We can influence the future but not the past. In other words, backward causation is impossible, de facto nonexistent, or, at the very least, extremely rare in this part of the world. For even if backward causation *sometimes* occurs, it remains to be explained why *predominant* direction of causation is toward the future. (7) Explanation: We seek explanations of phenomena in terms of antecedent, rather than subsequent, circumstances. I think that our ideas about the proper direction of explanation are intimately bound up with the directionality of cause and effect. What is not so clear, however, is precisely these phenomena are related. A natural answer is that *explaining* is a matter of *specifying causes* and consequently that the explanation of asymmetry is a product of causal directionality. (8) Counterfactual dependence: If the present were different from the way, it is, then the future would be different. Thus, true counterfactual conditional statements of the form "if A had not occurred, then C would not have" seem to be about what would have happened subsequently if some actual event had not taken place, and no about what would have happened before. Causation is a closely affiliated notion. David Lewis argues that counterfactual dependence *is* time-asymmetric and that the direction of causation is an immediate consequence of this phenomenon. (9) Decision: We act for the sake of the future, not the past. We think it rational to act only for the sake of things we might *cause* to occur, and we understand events, including our own actions, can exhibit causal influence only over the future. (10) Value: We care a great deal more about what *will* happen to us than about what *has* happened. Consequently, we would much rather have pleasant prospects ahead and bad times behind us the other way around.[12]

[12] Paul Horwich (1992: 4-11).

III

Phenomenological Chronometry

Time and chronicle can define thanks to the game of two of three clock needles; each position is taken maybe once inside a certain duration as the meteorological registrations. The singularity of each instant is bound to a stronger principle that the identity of the indiscernibles because what distinguishes to an instant *t* of an instant *t'* is not maybe that there is an event *p* that takes places in *t* and not in *t'* but rather *t* also be distinguished of *t''* for the presence of an event *q*, of *t'''* for *r*, etc. We can go farther until thinking that exists in each instant an event (clock event postulate *Ct*) that doesn´t find at any other instant again $(\forall t)(\exists p)(\forall t')[Tt'(p) \equiv (t=t')]$, i. e. whichever it is the instant *t*, an event *p* exists (characteristic of that instant) such that for all *t'*, *p* is not in *t'* again, provided that *t* and *t'* are not distinguished. The idea is that instants are abstractions of the events in such a way that the former can be distinguished based on the chronicles that happen in them. In practice, such events behave as a watchmaking mechanism. Therefore, we can distinguish the twelve on the dot of one o´clock for the chronicles in the positions of the pointers of the clock. An "event-clock" is a chronicle that fixes a time, assuming the occurrence that such chronicle in particular will be determined. The postulates of the event-clock is the stipulation that for each moment of time there is an "event-clock" that is given in this time and only in this[13]. We assume discreetness. If the discreteness coincides with the succession of the natural numbers, the five axioms of Peano:

(1) 0 is a number.
(2) The successor of all numbers is a number.
(3) There are not two numbers that have the same successor.
(4) 0 is not successor of any number.
(5) All classes that contain 0 and that contains the successor of *n* whenever that contains *n*, it includes the class of the numbers.

They become five axioms of temporal succession:

(1) t_0, is an instant of time.
(2) The successor of an instant of time is another instant of time.
(3) There are not two instants of time that have the same successor.
(4) t_0, is not successor of any temporal instant.

---

13 Rescher y Urquhart (1971: 176). See also Jean- Louis Gardies (1979: 109).

(5) All classes that contain t_0, and that contains the successor of t_n, whenever it contains n, it includes the class of all the instants of time.

We shall write C_t(p) to indicate that p is a clock-event proposition for the time t, i.e.:

$C_t(\text{p}) \equiv (\forall t')[\text{T}t'(\text{p}) \equiv (t=t')]$[14]

and these five axioms of temporal succession become five axioms C_t:

(1) C_t es un clock-event.
(2) The successor de C_t is another C_t.
(3) There is no two C_t that have the same successor.
(4) C_{t0}, is not successor of any instant.
(5) All classes that contain C_{t0}, and that contain the successor of C_{tn}, whenever contains n, it includes the class of all C_t.

Thus, the clock axiom amounts to a universal principle for clock-event propositions:

$(\forall t)\ (\exists p) C_t(p)$

We shall obviously have such rules as:

$C_t(p) \wedge C_t(q) \supset C_t(p \wedge q)$
$C_t(p) \supset Tt(p)$

It is crucial to distinguish $\sim C_t(p)$ y $C_t(\sim p)$. The former amounts to

$\sim(\forall t')[Tt'(p) \equiv t=t']$, or, equivalently

$(\exists t')\sim[Tt'(p) \equiv t=t']$

The second, on the other hand, comes to

$(\forall t')[Tt'(\sim p) \equiv t=t']$, or, equivalently

$(\forall t')[\sim Tt'(p) \equiv t=t']$, which implies the above but is not entailed by it.

It is worth that, on the relative conception of time, the clock-event postulate could be strengthened to:

---

[14] Rescher y Urquhart (1971: 1976).

$(\forall t)(\exists \underline{p})(\forall t')[Tt'(\underline{p}) \equiv (t=t')]$

The use of underscore is indicating as usual that the proposition in question is temporally pure or purely phenomenological. In this form the axiom leads to the consequence

$(\forall t)\ (\exists \underline{p})C_t(\underline{p})$

Which clearly expresses the central tenet of a theory of relative time that times are to be individuated phenomenologically on the basis of the events occurring "at" them[15].

Nicholas Rescher takes a Step Forward

Closely allied to the clock-event postulate is a chronometric axiom dealing with the proposition $C_t(p)$ for: "The clock now reads time t". We shall specify this by a contextual rather than direct definition

$Tt'[C(t) \equiv (t=t')]$

More fully, the *C*-operator is embodied in the full statement of a certain chronometric axiom:

$(\forall t)(\forall t')\{Tt'[C(t) \equiv t=t'\}$

The question –point out Rescher- of the existence of this *C*-operator amounts to the acceptability of:

$(\forall t)(\forall t')\ (\exists p)\{Tt'(p) \equiv t=t'\}$

But this is an immediate consequence of the clock-event postulate.

Now, Rescher goes beyond:

$Tt'\ [C(t) \wedge p] \equiv [t=t' \wedge Tt(p)]$

Rescher rewrites the Megarian omnitemporal modalities

$LA \equiv (\forall t)Tt(A)$
$MA \equiv (\exists t)Tt(A)$, for establish the theorem:

---

[15] Rescher y Urquhart (1971: 177).

$M[C(t) \wedge p] \equiv Tt(p)$

Thus, any proposition that is temporally compatible (in the Megarian manner) with $C(t)$ must be realized at t, and conversely

Again, Rescher established the theorem:

$MC(t) \wedge M{\sim}C(t)$

The philosopher of Pittsburgh University establishes to the Megarian modalities the two propositions:

(1) $M[C(t) \wedge p] \equiv Tt(p)$
(2) $MC(t) \wedge M{\sim}C(t)$

Finally, we arrive where I want, if recall the two axioms of Meredith (in the logic of world states), namely

(a) $Mw \wedge M{\sim}w$
(b) $M(w \wedge p) \supset p$, and until the last minute:

If we take the special case of $t = t'$, then $C(t) \equiv C(n)$ will play exactly the role of Meredith´s w, since (2) yields (a) y (1) now comes to:

$M[C(n) \wedge p] \equiv Tn(p) \equiv p$

And in fact, is possible to go beyond this when we temporalize w to obtain w_t (with $w_n = w$) subject to the rules

(i) $Mw_t \wedge {\sim}w_t$
(ii) $M(w_t \wedge p) \supset Tt(p)$

We have proved that $C(t)$ plays exactly the role of this temporally relativized world operator w_t[16].

We can coordinate the logic of worlds with the phenomenological chronometry.

---

[16] Rescher y Urquhart (1971: 178-179).

IV

One Tell

Gordon Childe invites us to practice speleology. The speleology is the science that studies the formation of caverns, flora and fauna. The logical of the time becomes speleologist. The misleading risk is always present. We need a map of time. If we dive in the geological memory, do we wonder what it means that instantaneous of 18 meters?

To obtain some idea of archaeological time let us consider the ruins of cities in Mesopotamia. The dead level of the alluvial plain between the Tigris and Euphrates is broken by *tells* or mounds, rising 60 feet or more above the surroundings ground. They are not natural hillocks, but each marks the site of ancient settlement, and is composed entirely of the debris of ruined houses, temples and palaces. In Iraq houses are still build of mudbricks, not baked in a kiln, but simply dried in the sun. Such a house may with luck stand for a century. But eventually the rain will penetrate under the eaves or reach the foundations and disintegrate the plastic mud. The whole edifice will then flop down, a shapeless mass of mud or crumbling dust. The owner will not bother to cart away the debris. He will just level it down and build on the old site a new house with its foundations a couple of feet higher than the floor of the previous dwelling. The repetition of this process over successive centuries is responsible for the *tells* that break the monotony of the Mesopotamian plain.

At Warka, the biblical Erech, the Germans explored the center of such a *tell* by means of a deep shaft. The top of the shaft is the floor level of a temple that itself prehistoric and about 5500 years old. From his level you may described by winding path cut in the walls of the shaft for a depth over 60 feet! At every stage of the dizzy descent, you can pick from the shaft´s sides bits of pottery, mud bricks, or stone implements. The shaft is actually cutting a mound, 60 feet high composed entirely of the debris of successive settlements in which men had lived. The mound has grown up in the manner described above, but even the latest of the component settlements traversed in descending the shaft is more than five millennia old!

At the bottom we reach virgin soil –the soil of marsh that just emerged from the Persian Gulf. The lowest settlement represents the remote beginning of human life in Southern Mesopotamia. But when we have descended to it, we are as far as ever from the beginning of human progress. To achieve that we must plunge into geological time. But now figures become almost meaningless (and mainly guesswork). To grasp man´s antiquity we must consider the vast changes in the earth´s surface that our species had witnessed before ever the first settlers reached the site of Erech[17].

We need, of course, a registration book, a speleology notebook or a watch.

[17] Gordon Childe (1996: 59).

V

The Clock Axiom

In the Appendix B of *Past, Present and Future,* Section 9, titled "Anticipations of later developments in Łoś Calculus of Instants"[18], Arthur Prior remember Jerzy Łoś, the calculus which devised in 1947 in an attempt to formalize Mill's canons of inductions. The calculus appeared in the *Annales Universitatis Mariae Curie-Sklodowska,* Section F, Vol. 2 (for 1847, published in 1948), pp. 269-301, and was summarized and reviewed by Henry Hiż in the *Journal of Symbolic Logic,* vol. 16. No. 1 (March 1951), pp. 58-59. Łoś calculus has

- No tense operators, but does use propositional variables p_1, p_2, etc., to stand for what might be "satisfied" at one instant and not another.
- Variables t_1, t_2, to stand for instants and n_1, n_2, etc., for temporal intervals.
- The form Ut_1p_1 for "p_1" is satisfied at t_1, and $\delta t_1 n_1$ for "the instant n_1 later than t_1. He abridges $(\forall p_1)(Ut_1p_1 \equiv Ut_2p_2)$ to pt_1t_2, which may be read as "t_1 and t_2 are the same instant".

Prior recognizes that this calculus influenced his own formulation of a

"Calculus of dates" (using the form *Utp*) in *Time and Modality and* has points of resemblance to Rescher system of 1965. It will facilitate comparisons if we give Łoś's axioms in the symbolism of Chapter 6, Section 4, supplemented by *San* for "the instant *n* later than *a*", and "*a=b* is the same instant". (a=b, it should be remembered, is short for $(\forall p)(Tap \equiv Tbp)$[19] and *TSanp* is equivalent to *TaFnp* in the symbolism of Chapter 6, Section 4). The axioms then become:

1. $Ta{\sim}p \equiv {\sim}Tap$
2. $Ta(p \supset q) \supset (Tap \supset Taq)$

---

[18] Prior (1967: 212-213).

[19] I should remember that *Tap* to mean "It is the case at the date *a* that *p*". Axiomatically the logic of dates:
TN1: $Ta{\sim}p \supset {\sim}Tap$; TN2: ${\sim}Tap \supset Ta{\sim}p$; T∀1: $(\forall a)(Tap \supset p)$;
T∀2: $(\forall b)(Tbp \supset Ta(\forall b)Tbp)$;
TC: $Ta(p \supset q) \supset (Tap \supset Taq)$; TT: $TaTbp \supset Tbp$. If we use *F and P* in which interval-variables stand for positive numbers , but allow date variables to stand for positive, negative, and zero numbers, the mixing laws are: TF: $TaFnp \supset T(a+n)p$; TP: $TaPnp \supset T(a-n)p$; FT: $FnTap \supset Tap$; PT: $PnTap \supset Tap$. The converses of these are derivable.

3, 4 and 5. $Ta(p\supset q)\supset[(q\supset r)\supset(p\supset r)]$, $Ta(p\supset(\supset p\supset q))$, $Ta(\sim p\supset p)\supset p$

6. $(\forall a)Ta(p\supset p)$

7 and 8. $(\exists b)San=b$, $(\exists b)Sbn=a$

9. $(\exists p)(\forall b)(Tbp\equiv a=b)$

- The axioms 1-5 have as consequences all propositional-calculus theorems preceded by *Ta*. (cf. the rule, in Rescher's systems and mine, to infer $\vdash Ta\alpha$ from $\vdash\alpha$.

- 7 would seem to be replaceable by the permission to substitute any expressions of the form *San* for instant-variables in theses. (…) Łoś apparently thought that this requires "that there be an infinite number of constants which can be substituted for the variables representing instants"; Hiż argues in his review that this would only be the case if we had an axiom, say $San=b\supset a\neq b$, *excluding circularity.*

- Axiom 9, the "clock axiom", asserts in effect that "to every instant of time a function can be assigned (e. g. the description of the position of the hands of a clock) which is satisfied only at that instant". Łoś regarded it as "our only weapon against the metaphysical and extrasensual conception of time". His point of view seems in fact to have been very close to Sections 3 and 4 of this appendix [Section 3:" On the range of worlds-variables, and the interpretation of U-calculi in world-calculi"; Section 4: "The uniqueness of the time-series".] The clock axiom, one might say, might (or might reflect) our *identification* of an "instant" with a proposition true at that instant only; it corresponds to the postulates *Taa* y $Tab\supset a=b$ in a system using instant variables as a special sub-class of propositional ones.

The resemblance with Rescher's clock-event postulate is clear.

VI

The Models Come to the Fore

Ptolemy built a model of the universe postulating the geocentrism, the immobility of the earth and a complicated epicycles system to reduce the apparent movement of the planets regarding the fixed stars; Copernicus, other completely different, he maintained the heliocentrism and the circular movement of the planets around the sun; Copernicus reduced from 79 to 34 the epicycles number and he believed, the same of Ptolemy that the planets were connected with others as pearls of a necklace. A model seeks to be an isomorphic representation of reality. A Model was the structure of double helix thar that built James Watson and Francis Crick in 1953 for the DNA like the basic component of genetic material. A model is the atomic structure fixed by Rutherford and Bohr at the beginning of XX century. A model was the discovery for Newton of the law of gravity, unifying this way the terrestrial physics with the celestial one. A model was also the theory of the special relativity by Einstein. Wittgenstein in the proposition 2.1 of *Tractatus*[20]. advises: “We picture facts to ourselves”, also, in the proposition 2.12 “A picture is a model of reality”. Finally, in the proposition 2.141, the Viennese philosopher defines “picture” like: “A picture is a fact”. Bas C. Van Fraassen in *The Scientific Image*, distinguishes between what calls the “semantics” as opposed to the syntactic approach to the theories. Traditionally in the history of philosophy and of mathematics, a theory is considered as a body of propositions or statements, expressed in a syntax that can be axiomatized or alternatively it can be demonstrated in a deductive way (e.g. the Modus Ponens). But there is another way to see a theory. A theory can be considered *more semanticae*, like a collection of models or structures that, given a convenient interpretation, they make true to a set of propositions:

la representación sintáctica de una teoría identifica a ésta con un conjunto de teoremas, expuestos en un lenguaje particular elegido para su expresión. Esto debe ser comparado con la alternativa de presentar una teoría, a primera instancia, a través de la identificación de una clase de estructuras como sus modelos. Desde un punto de vista semántico, el lenguaje usado para expresar la teoría no es ni básico ni único; la misma clase de estructuras puede muy bien ser descrita en sentidos radicalmente diferentes, cada una con sus propias limitaciones. Los modelos ocupan el centro de la escena[21].

[20] Wittgenstein *Tractatus Logico-Philosophicus* (1921).

[21] Bas C. Van Fraassen (1980: 41).

Wittgenstein conceives the pattern of world like the logical pictures in the proposition 2.19: "Logical pictures can depict the world".

Number 1 proposition of *Tractatus* notes:

> 1. The world is all, that is the case.

At the same time, number 2 defines a fact:

> 2. What is the case -a fact- is the existence of states of affairs.

2.04, defines the world:

> 2.04. The totality of existing states is the world.

Wittgenstein distinguishes between reality and world at 2.06 y 2.063:

> 2.06. The existence and non-existence of states of affairs is reality.
> 2.063. The sum-total of reality is the world.

VII

The Game is World

The game is called "the world". Each movement is registered in the world like an index of instant. Each world-proposition is in each instant[22]:

A world-state proposition in the tense-logical is simply an *index of an instant,* I would like to say that it is an instant, in the only sense in which "instant" are not highly fictions entities. To be the case at such-and-such an instant is simply to be the case in such-and-such a world; and that in turn is simply to be the case *when* such-and-such a world-proposition is the case. In this sense of "instant" it is a tautology that a world proposition is true at one instant only (it is true only when *that* world-proposition is true) and so is as "strong" as any proposition that is ever true can be; though if time is circular it will not follow from his that if a world-proposition *is* true it neither has been nor will be true (for in circular time, "has been", and "will be" bring us back eventually to *the same* instant).

Where can I write this *index of instants* in my book of speleology? Prior declare: "in the domain of the relation *U*" but

the axioms A1 [Ma] y A2 [$L(a\supset p)\vee(a\supset\sim p)$] of Chapter V, Section 6, do not quite do this. A1, Ma secures that each world-proposition is true at some time or other; but there ought also to have been a postulate securing that every point of time has a world-state which "occupies" it (or which it is). What we want here is simply $(\exists a)a$, from which we derive $L(\exists a)a$ by RL. We may call $(\exists a)a$ A3, but in what follows we shall only be using a derivative of it, asserting that if anything is ever true there is some world-state "in" which it is true, i.e. which permanently implies it: $Mp\sim(\exists a)L(a\supset p)$. We may call this A3'. In deriving it from A3 we use the Barcan Formula (BF) for worlds, which we know we have when given at least the system B for *L and M.* Prior also use the quantification theory,

We also make use of theorem of quantification theory that although we do not have $(\exists x)\Phi x\wedge(\exists x)\psi x\supset(\exists x)(\Phi x\wedge\psi x)$, we do have $(\exists x)(\Phi x\wedge p)\supset(\exists x)(\Phi x\wedge p)$. We cannot expect to derive a *U-calculus* within the corresponding tense-logic plus the calculus of worlds, unless the latter

[22] Prior (1967: 188, 189), Prior (1968: 117-134), Peter Øhrstrom and Per F. V. Hasle (1995: 219-229). Rescher y Urquhart (971: 179).

A3 added to it. (A3 is needed, we shall find, for the derivation of the right-to-left implication in U3 y U4[23]).

Finally, even A3, we cannot expect to do it in any tense logic system but one for linear (non-forking) time, if we define $M\alpha$ as $\alpha \vee P\alpha \vee F\alpha$ y $L\alpha$ as $\alpha \wedge H\alpha \wedge G\alpha$. I have found it simplest to make this point clear to myself by associating *U*-calculi with linear diagrams. In the world-calculus within which (together with some tense-logic) we wish to develop the *U*-calculus, we want *Lp* to mean in effect that *p* is true *all over the diagram,* and *Mp* that *p* is true *somewhere in the diagram.* And it is only in linear time that *p* is true-somewhere-on-the-diagram if and only if it either is true now, or has been true, or will be true.

We Consider the diagram:

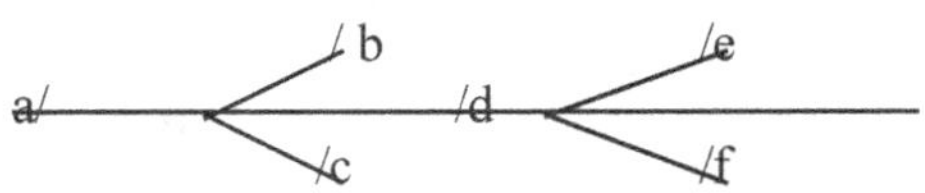

If world *d* is what is the case now, then each of *a, d, e*, and *f*, satisfy this condition "It either is true now (like *d*) or has been true (like *a*) or in some future will be true (like *e* and *f*)". But, the "could-have-beens" *b* and *c not* satisfy this condition, although they are "true-somewhere-on-the diagram" [24]. Further, we can only be sure that even *a, e,* and *f*, satisfy the condition if earlier-later relation is taken to be transitive. Transitiveness seems, indeed, to be assumed in any representation of futurity by being "somewhere" to the right and pastness by being "somewhere" to the left".

In short, the inference tickets par excellence is the transitiveness and linear time[25].

---

[23] U1. $Ta{\sim}p \equiv {\sim}Tap$; U2. $Ta(p \supset q) \equiv (Tap \supset Taq)$; U3. $TaGp \equiv (\forall b)(Uab \supset Tbp)$; U4. $TaHp \equiv (\forall b)(Uba \supset Tbp)$; see, Prior, "Development of the *U*-Calculus within the theory of world-states", *op. cit.*, 1967, p. 88.

[24] Prior (1967: 190).

[25] For concept "inference ticket" see also Storrs McCall (1994: 1966) "*Space-Time, Probability, and Decision*" send us a Ryle (1950) "If, So, and Because" in Max Black (ed.), *Philosophical Analysis* (Ithaca, NY), pp. 302-318, esp. 308-309. Ryle support that conditionals are statements with truth-values.

VIII

The Domain of *U*

Prior asserts that $Lp=(\forall n)L^n p$ and $Mp=(\exists n)\ L^n p$, again will make the postulate $M\alpha$ (A1)

assert that every term in the field of the relation U is in *some* way temporally connected with the present world. If, on the other hand, L is *defined* as $(\forall n)L^n$, and is regarded as only intelligible in some such terms, there is just no alternative to this equivalence[26].

What formal consequences are defined as $Lp=(\forall n)L^n p$?

(1) If we subjoin $Gp\supset GGp$, $p\wedge Hp\wedge Gp\supset HGp$, and $p\wedge Hp\wedge Gp\supset GHp$ (i. e. the tense-logical expressions of transitivity and non-branching both ways) to K_t, Lp $(=p\wedge Hp\wedge Gp)$ becomes equivalent to Lp; that

(2) if we add the first and just one of the other two, $L^2=L$; *and that*

(3) if we add $Gp\supset GGp$, $Gp\supset Hp$, *and* $Gp\supset p$ (postulates for circularity), even the plain $G=L$[27].

Let us observe (1):

It was remarked by Lemmon in 1965 that there seems to be no purely tense-logical formula which corresponds exactly to the U-condition $Uab\vee Uba\vee a=b$, which he calls strict or strong linearity. There are indeed tense-logical formulae, e. g. $p\wedge Hp\wedge Gp\supset HGp$, and $p\wedge Hp\wedge Gp\supset GHp$ which corresponds exactly to non-branching in both directions, i. e. to the pair of condition:

$Uab\supset(Uac\supset Ubc\vee Ucb\vee b=c)$ [linearity to future]
$Uba\supset(Uca\supset Ubc\vee Ucb\ \vee b=c)$ [linearity to past],

Respectively[28].

Dialoguing with Lemmon, Prior pointed out:

---

[26] Prior (1967: 198).
[27] Prior (1967: 201).
[28] Prior (1967: 201).

But, Lemmon pointed out, these conditions and these formulae are compatible with there being several unconnected times-series each of which is separately linear; the categorical $Uab \vee Uba \vee a=b$, is not compatible with this, and it is this exclusion of the possibility of distinct times-series which no purely tense-logical formula seems to capture. But we do capture it by the formula proved above $p \wedge Hp \wedge Gp \supset Lp$, which says in effect that if *p* is and always has been and always will be true, it is true in all worlds there are. This intuitively exclude a plurality of time-series and can be used formally to prove the strict linearity condition on *U*. (...) The truth of Lemmon's contention thus depends on whether $p \wedge Hp \wedge Gp \supset Lp$, is a purely tense-logical formula or not. If *L* is not tense-logically defined, it is not, and Lemmon's contention stands; but if it is defined as $\forall nL^n$, it is, and a plurality of time-series is tense-logically excluded[29].

Prior obtains identical results with the circular time:

Circularity in the sense of the transitivity, symmetry, and reflexiveness of *U* does not in itself preclude there being a number of distinct circular time-series; for this we need $\vdash Uab$, i. e. "*Every* world is earlier than (and later than) every other"; and we could obtain this if we had $Gp \supset Lp$ from the usual circularity axioms if we define *L as* $\forall nL^n$, but if we take it as undefined (with postulates S5+$Lp \supset Hp$+$Lp \supset Gp$), we do not.
In a *U*-calculus which is not tense-logically anchored, but is taken as basic, and in which tensed formulae (and formulae with *L*) occur only as second argument of the functor *T* (i. e. in which the *tensed* proposition α is replaced by tensed *predicate* $T\alpha$), some of the last-mentioned results may be presented as follows: To the usual basic equivalences *U1-U4* [see note 10], we add

U5. $TaLp \equiv (\forall b) Tbp$[30]

Finally, Prior anticipates the Four Grade of Tense-Logical Involvement[31]

It is well known that with this for *L* we have all S5 theses preceded by *Ta*; and it is easy prove $Ta(Lp \supset Gp)$ y $Ta(Lp \supset Hp)$. The converses $Ta(Gp \supset Lp)$ and $Ta(Hp \supset Lp)$ are provable if we add $\vdash Uab$ to our basis, i. e. if we equate *U* with the universal relation; and $Ta(p \wedge Gp \wedge Hp \supset Lp)$ if we lay down $\vdash Uab \vee Uba \vee a=b$, i. e. equate $(U \cup \breve{U} \cup I)$, the logical sum of U, its converse

[29] Prior (1967: 202).
[30] Prior (1967: 203).
[31] Prior (1968: 116-144).

and identity, with the universal relation. Unconnected *U*-series would be excluded by laying it down that the ancestral of the logical sum *U* an its converse and identity relates every *a* and *b*, i. e. $\vdash (U \cup \breve{U} \cup I)^{*}ab$, or $\vdash (U \cup \breve{U} \cup I)^{*} = V$[32]. Given this, we may prove

$\vdash Uab$ from *U's* being transitive, reflexive, and symmetrical, and

$\vdash Uab \vee Uba \vee a=b$, from its being transitive and non-divergent both ways.

[32] Prior (1968: 163).

IX

Seeds of All Possible Worlds (homoioméreia of worlds): The Function *Wp*

Again, we don´t abandon Vienna, the proposition 6.13: "Logic is not a body of doctrine, but a mirror-image of the world. Logic is "the great mirror". Arthur Prior note in *Past, Present and Future*, Chapter V, 1. "The de-trivializing of modality"

C.A. Meredith has suggested[33] that one logically interesting contingent proposition is "the world" as defined in the first sentence of Wittgenstein´s *Tractatus* – "Everything that is the case". For this "sum of all truth", Meredith introduces the symbol n, with the axioms

1.	n	("The world is the case")
2.	$p \supset L(n \supset p)$	("The world is everything that is the case")
3.	$Ln \supset p$	("The world is not necessary").

Literally, 2 says that if it is the case that p, then "The world" necessarily implies that p; and 3 that if "the world" is necessary, anything at all is the case.

Meredith´s variant makes it possible to define *L* in terms of *n*.

Meredith therefore subjoins it to a modal logic which does not have this rule, though it contains the same theses in ⊃, ~, and as Lewis´s S5. This calculus does not survive the translation of *Lp* as p, and for consistency Meredith gives a 4-valued matrix, which can be interpreted by supposing that there are only two possible worlds, n (the actual one) and ñ, and the four "values" that propositions can take are:

- "true in both worlds" (i.e. necessarily),
- "true in the actual world only" (i.e. contingently true),
- "true in the alternative world only" (i.e. contingently false) and
- "false in both worlds" (i.e. necessarily false),

and the laws are those formulae which always hold (i.e. for all values of their variables) either in this world only or in both. The matrix is

[33] C. A. Meredith and A. N. Prior (April 1965: 99-109).

⊃	1	n	ñ	0	L
*1	1	n	ñ	0	1
*n	1	1	ñ	ñ	0
ñ	1	n	1	n	0
o	1	1	1	1	0

Prior indicates that it was R. Suszko who pointed out that this solution to the problem of confusion between *Lp*, *Mp*, and *p* can be by dropping Meredith´s constant, and introducing instead a function *Wp* which asserts

that *p* has the properties of that constant, i.e. that *p* is a truth so comprehensive that all other truths follow from it.

We can define *Wp* as $p\wedge(\forall q)[q\supset L(p\supset q)]$, "*p* is a truth, and for any *q*, if *q* is true then *p* necessarily implies it". We can obtain the Meredith´s axioms under a condition, i.e.

1. $Wp\supset p$

And

2. $Wp\supset[q\supset L(p\supset q)]$.

For the third, we need to lay down that there is at least one contingent truth,

3. $(\exists p)(p\wedge M{\sim}p)$,

from which it follows that the totality of truth is contingent

4. $Wp\supset M{\sim}p$ (o $Wp\supset{\sim}Lp$, o $Wp\supset L(p\supset q)$).

The newness is that we can lay down

5. $(\exists p)Wp$ as a further axiom.

Finally, if *p* and *q* are both all-comprehensive truths, they are necessarily equivalent,

6. $Wp\supset[Wq\supset L(p\equiv q)]$.

X

The Footnote

In his paper Findlay wrote:

Before an event occurs we say, if we have evidence that it is not yet happening, that it hasn't yet happening, but that it will happen, while if it *is* happening we say that it is now happening, that it hasn't ceased happening and that it isn't about happen, and *after* it has happened we say that it has happened, that it is no longer happening, and that it is not going to happen. Stated in words these semantic rules might seem circular but taught in connection with a concrete situation they are wholly clear. And our conventions about tenses are so well worked out that we have practically the materials in them for a formal calculus.

The following footnote:

The calculus of tenses should have been included in the modern development of modal logics. It includes such obvious propositions as that:

x present = (x present) present;

x future = (x future) present = (x present) future;

also, such comparatively recondite propositions as that

(x).(x past) future; i.e. all events, past, present, and future, *will* be past[34].

We formalize this result:

The symbol ≡ is the logic equivalence ↔.

J = ahora que equivale al presente.

x futuro = Fx

x pasado = Px

Jx = (Jx)Jx

Fx = J(Fx) = F(Jx)

$Jx \vee Px \vee Fx \supset FPx$, all events, past, present, and future, *will* be past.

[34] J. N. Findlay (1941: 143-162).

XI

The Prior's J

Prior devised the J system, i.e. the system of now[35]: "Suppose we now write *Jp* for "it is *now* the case that *p*". For this operator we do need the following:

J1. $p \supset Jp$
J2. $Jp \supset p$

Prior starts by incorporating *n* of Meredith. The new symbol gives a new axiom, viz

A3. n[36].

and from the UT Calculus the variables a, b, c, etc., may stand for any proposition true at one instant and at one instant only. For these we have the two axioms

A4. $\sim L \sim a$
A5. $L(a \supset p) \vee L(a \supset \sim p)$.

Prior bring out that using Mp ("p at some time") as short for ~L~a, we can abridge A4 to *Ma*. And then Prior indicates, "We define Jp either as $M(n \wedge p)$ ('At some time, p true together NOW'), or as $L(n \supset p)$ ('p true at all times at which NOW'), alternatives which the uniqueness of NOW makes equivalent; and we enlarge the restriction on RL to 'provided that α does not contain J or n'". The postulates RL are:

RL: $\vdash \alpha \rightarrow \vdash L\alpha$, provided that α does not contain J.
L1. $Lp \supset p$
L2. $L(p \supset q) \supset (Lp \supset Lq)$
L3. $\sim Lp \supset L \sim Lp$
L4. $Lp \supset Gp$
L5. $Lp \supset Hp$.

---

35 *PTT* "Now" (2003: 171-193).

36 The A1s and A2s axioms:

A1.1. $G(p \supset q) \supset (Gp \supset Gq)$ A2.1. $PGp \supset p$
A1.2. $H(p \supset q) \supset (Hp \supset Hq)$ A2.2. $FHp \supset p$
PTT (2003: 177).

Prior bring out that

(i) RL and L1-3 (subjoined to propositional calculus) are known to suffice for the pure *L* portion of the required calculus, which is the modal system S5[37].

The postulates RL (modified), L1-5 and A1.1-A5, with the definition of *J*, will I think yield all the theorems in *J* that we want; and the definition of "now" as expressing contemporaneity[38]. (...)

But Prior suggests drop A3-5 and the proposed definition of J and give postulates for an undefined J to be added to RL, L1-5 and A1.1-A2.2 of Kt, namely the following:

J1. $p \supset Jp$
J2. $Jp \supset p$
J3. $L(Lp \supset Jp)$
J4. $L(Jp \supset LJp)$
J5. $L(J{\sim}p \supset {\sim}Jp)$
J6. $L({\sim}Jp \supset J{\sim}p)$
J7. $L[J(p \supset q) \supset (Jp \supset Jq)]$[39]

These postulates may be reduced to J1, J2, and J4. Some considerations for Prior:

(1) From RL, L1-3, and J1-7 we can prove all of the following permanent equivalences $L(JJp \equiv Jp)$, $L(LJp \equiv Jp)$, $L(Lp \equiv JLp)$, $L({\sim}Jp \equiv J{\sim}p)$, $L[(Jp \supset Jq) \equiv J(p \supset q)]$. These enable us within this sub-calculus (i.e. the part without *G* or *H*), when all the variables in a formula fall within the scope of an *L* or a *J*, to equate the whole with a formula in which the only *J* occurs right at the beginning and there we can strengthen it to *LJ* by J4. So, RL applies to formulae of this sort as well as to ones with no *Js* in them at all.
(2) Given RL, L1-3 and A3-5, all of J1-7 are deducible from the definition of *Jp* as $L(n \supset p)$, or from $L[Jp \equiv L(n \supset p)]$ and conversely $L[Jp \equiv L(n \supset p)]$ from J1-7.

---

[37] *PTT* (2003: 182).
[38] *PTT* (2003: 184).
[39] *PTT* (2003: 184).

XII

The Meredith's δ

According to Jack Copeland[40]

It was probably Łukasiewicz who first introduced Meredith to modal logic. The two worked to extend the use of δ-variables to modal calculi. The axiomatization given by Łukasiewicz in February (1953) contained an axiom that Meredith had proposed during their earlier investigation of non-modal propositional logic, $\delta p \rightarrow ((\delta{\sim}p \rightarrow p) \rightarrow q)$. Substituting "M" for the variable "δ" produces $Mp \rightarrow ((M{\sim}p \rightarrow p) \rightarrow q)$. Perhaps Łukasiewicz did not find this implausible, but Meredith sought improvement, devising a system with δ variables whose δ-free portion is S5. This work went unpublished at the time and was later incorporated into Meredith and Prior 1965.

Prior defines δ of Meredith in this way mentioned before: $Wp \supset M{\sim}p$ (o $Wp \supset {\sim}Lp$, o $Wp \supset L(p \supset q)$.

Conclusion: $\delta = Mp \wedge M{\sim}p$, of course, the definition of contingent, both true together but not false.

According to Prior[41], the δ system be composed of ⊃ and an impossible (or permanently false) constant *0* ($\sim p = (p \supset 0)$, as well as δ for one-place connectives. He has the laws for L, n, and J:

M1. $L[\delta[(p \supset 0) \supset (q \supset r)]] \supset \delta[(r \supset p) \supset (q \supset p)]]$	
M2. $Lp \supset [\delta(p \supset q) \supset \delta q]$	
M3. $\delta 0 \supset [\delta(0 \supset 0) \supset \delta Lp]$	
M4. n	= A3
M5. $p \supset (n \supset p)$	
M6. $p \supset Jp$	= J1
M7. $Jp \supset p$	= J2
M8. $\delta(p \supset q) \supset \delta(Jp \supset Jq)$	
M9. $\delta J0 \supset \delta 0$	
M10. $\delta 0 \supset [\delta(0 \supset 0) \supset \delta Jp]$.	

Prior bring out that M1-3 are equivalent to the whole of $S5 + L(p \equiv q) \supset (\delta p \supset \delta q)$. For n, Meredith has also $Ln \supset p$, "which denies that our is the

[40] Jack Copeland (2002: 114).

[41] *PTT* (2oo3: 188).

only possible world, or in tense logic that now is the only instant; we omit it as out of place in a *minimal* system. He merely mentions M6-10 at the end of the article as being true of Jp defined as L(n⊃p) but as group there is more interest to them than that. Given M1-5, they yield as well as being yielded by *L[Jp≡L(n⊃p)]* (or δJp⊃δL(n⊃p)); and given just M1-3, they yield and are yielded for our own J1-7".

XIII

The Proofs

From J1, J2 and J4, and the Brouwersche system Prior deduces everyone else:

1. p⊃LJp	J1, p⊃Jp; J4, Jp⊃LJp, p.c.
2. p⊃L(Lp⊃Jp)	1; Lp⊃L(q⊃p); from LJp to L(Lp⊃Jp), Strict implication paradox.
3. ~p⊃~MLp	MLp⊃p, transp.
4. ~p⊃L(Lp⊃Jp)	3, ~Mp⊃L(p⊃q); from ~MLp to L(Lp⊃Jp), Strict implication paradox.
5. L(Lp⊃Jp)	2, 4, p∨~p = J3, Q.E.D
6. MJp⊃MLJp	J4, Jp⊃LJp, from Kt L(p⊃q)⊃(Mp⊃Mq), to MJp⊃MLJp, p/Jp, q/LJp
7. MJp⊃Jp	6, MLp⊃p, from MLJp to Jp, p/Jp
8. ~Jp⊃~MJp	7, transp.
9. ~Jp⊃L~Jp	8, ~Mp⊃L~p
10. ~p⊃L~Jp	J2, 9; J2, Jp⊃p, in this case, ~Jp⊃~p in the antecedent of 10.
11. ~p⊃L(J~p⊃~Jp)	10, Lp⊃L(q⊃p), from L~Jp to L(J~p⊃~Jp), Strict implication paradox.
12. ~~p⊃L~J~p	10, ~ in antecedent and consequent.
13. ~~p⊃L(J~p⊃~Jp)	12, L~p⊃L(p⊃q), from L~J~p to L(J~p⊃~Jp) Strict implication paradox.
14. L(J~p⊃~Jp)	11, 13 p∨~~p = J5 Q.E.D.
15. ~p⊃LJ~p	1, p/~p.
16. ~p⊃L(~Jp⊃J~p)	15, Lp⊃L(q⊃p), from LJ~p to L(~Jp⊃J~p), Strict implication paradox.
17. p⊃L(~Jp⊃J~p)	1, Lp⊃L(~p⊃q), if in (1) we have ~p⊃LJ~p, then from LJ~p to L(~Jp⊃J~p) by Strict Implication paradox.
18. L(~Jp⊃J~p)	16, 17, ~p∨p = J6, Q.E.D.
19. q⊃L[(J(p⊃q)⊃(Jp⊃Jq)]	1, Lp⊃L(q⊃p), two times; 1 = p⊃LJp; from LJp to L(Jq⊃Jq) by Strict implication paradox, and then Jq⊃Jq is implied by J(q⊃p), hence,

L[(J(q⊃p)⊃(Jq⊃Jp)], and then is implied by q, but p/q and q/p.

20. MJp⊃p 7, J2, p.c.

21. [MJ(p⊃q)∧MJp]⊃q [(p⊃q)∧p]⊃q, 20.

22. M[J(p⊃q)∧Jp]⊃q 21, M(p∧q)⊃(Mp∧Mq) theorem from Kt

23. ~q⊃~M[J(p⊃q)∧Jp] 22, transp.

24. ~q⊃L[J(p⊃q)⊃~Jp] 23, ~M(p∧q)⊃L(p⊃~q), in the consequent

25. ~q⊃L[J(p⊃q)⊃(Jp⊃Jq)] 24, ~p⊃(p⊃q), from ~Jp to (Jp⊃Jq) by Strict implication paradox.

26. L[J(p⊃q)⊃(Jp⊃Jq)] 19, 25, q∨~q = J7, Q.E.D[42]

---

[42] *PTT* (2003: 189, 190).

XIV

The Folder condition

Folder and unfolder condition are a Metaphysics concept about time and reality based in the Prior´s J and the Meredith´s δ. Prior proves the equivalences from RL, L1-7, J1-7[43]

1. LJ(p⊃Jp)	J1, J4; J1 = p⊃Jp, J4 = L(Jp⊃LJp), p/p⊃Jp.
2. L(Jp⊃JJp)	L2, J7, 1; J7 = L[J(p⊃q)⊃(Jp⊃Jq)], then from LJ(p⊃Jp), to L(Jp⊃JJp).
3. LJ(Jp⊃p)	J2, J1, J4; such as 1.
4. L(JJp⊃Jp)	L2, J7, 3.
5. L(JJp≡Jp)	2, 4; p.c.
6. L(Lp⊃p)	L1, RL.
7. L(LJp≡Jp)	6, p/Jp, J4; J4 = L(Jp⊃LJp).
8. L(~Lp⊃J~Lp)	L3, RL, J3; L3 = ~Lp⊃L~Lp, J3 = L(Lp⊃Jp); then, ~Lp⊃J~Lp and RL.
9. L(~Lp⊃~JLp)	8, J5; J5 = L(J~p⊃~Jp), p/Lp in consequent
10. L(JLp⊃Lp)	9, transp.
11. L(LLp⊃JLp)	J3, p/Lp.
12. L(Lp⊃JLp)	11, L(Lp⊃LLp) from S5.
13. L(Lp≡JLp)	10, 12; p.c.
14. L(~Jp≡J~p)	J5, J6; J5 = L(J~p⊃~Jp) and J6 = L(~Jp⊃J~p).
15. L[J~(p⊃q)⊃J(p∧~q)]	~(p⊃q)⊃(p∧~q), J1, J4, J7; Ok, ~(p⊃q)⊃(p∧~q), and then J~(p⊃q)⊃J(p∧~q) by J7, finally, by J4, L[J~(p⊃q)⊃J(p∧~q)].
16. L[J(p∧~q)⊃Jp]	p∧q⊃p, J1, J4, J7; we see it, we have J(p∧~q)⊃Jp, p/p∧~q, q/p, and then by J4, L[J(p∧~q)⊃Jp].
17. L[J(p∧~q)⊃J~q]	p∧q⊃q, J1, J4, J7, such as 16.
18. L[J(p∧~q)⊃~Jq]	17, J5; by J5, J~q = ~Jq, in consequent
19. L[J(p∧~q)⊃(Jp∧~Jq)]	16, 18; p.c.
20. L[~J(p⊃q)⊃~(Jp⊃Jq)	J6, 15, 19, L(p∧~q)⊃~(p⊃q) in antecedent and consequent

[43] *PTT* (2003: 191).

21. $L[(Jp\supset Jq)\supset J(p\supset q)]$ 20, transp.
22. $L[(Jp\supset Jq)\equiv J(p\supset q)]$ 21, J7.

We have, of course, the following equivalences:

5.- $L(JJp\equiv Jp)$
7.- $L(LJp\equiv Jp)$
13.- $L(Lp\equiv JLp)$
14.- $L(\sim Jp\equiv J\sim p)$
22.- $L[(Jp\supset Jq)\equiv J(p\supset q)]$

For the last equivalence, Prior adds A3, n, and A5, $L(a\supset p)\vee L(a\supset\sim p)$, to get[44]:

23. $L(n\supset p)\supset p$ — A3, $p\supset[L(p\supset q)\supset q]$.
24. $LJ[L(n\supset p)\supset p]$ — 23, J1, J4; we introduce LJ by J4.
25. $L[JL(n\supset p)\supset Jp]$ — L2, J7, 24; J7 = we introduce J in antecedent and consequent by J7.
26. $L[L(n\supset p)\supset Jp]$ — 25, 10; remove J in antecedent: $JL(n\supset p)\supset L(n\supset p)$, $p/n\supset p$.
27. $p\supset[\sim L(n\supset\sim p)]$ — 23, $p/\sim p$, transp. From $L(n\supset\sim p)\supset\sim p$ to $p\supset\sim L(n\supset\sim p)$.
28. $p\supset L(n\supset p)$ — 27, A5 = M5.
29. $L[Jp\supset L(n\supset p)]$ — 28, J1, J4, J7, 10; $Jp\supset L(n\supset p)$; second, $LJp\supset JL(n\supset p)$ by L2, J7; $LJp\supset L(n\supset p)$ by 10; and finally, $L[Jp\supset L(n\supset p)]$
30 $L[Jp\equiv L(n\supset p)]$ — 26, 29,

And next Prior resolves the following, adding M1-3 (from which Ext. $L(p\equiv q)\supset(\delta p\supset\delta q)$ and Df. ~ to get

31. $L[J(p\supset 0)\supset(Jp\supset 0)]$ — J5, Df. ~; J5: $L(J\sim p\supset\sim Jp)$ from the equivalence of $J(p\supset 0)$ to $J\sim p$ and $Jp\supset 0$ to $\sim Jp$
32. $LJ(0\supset 0)$ — $p\supset p$, J1, J4: $L(Jp\supset LJp)$, from $0\supset o$, false implies false.
33. $L(J0\supset 0)$ — 31, 32
34. $L(0\supset J0)$ — $0\supset p$, RL, we introduce L
35. $L(J0\equiv 0)$ — 33, 34

44 *PTT* (2003: *Ibid*).

36. δJ(p⊃q)⊃δ(Jp⊃Jq)	J7, Ext. = M8; first *J* and second δ
37. δJ0⊃δ0	35, Ext.=M9; we apply L(p≡q)⊃(δp⊃δq) in 35.
38. δLJp⊃δJp	7, Ext. 7: LJp⊃Jp. We apply δ in Antecedent and consequent.
39. δ0⊃[δ(0⊃0)⊃δJp]	M3, p/Jp, 38 = M10, if δ is false, then the implication false of δ implies δJp[45]

Then Prior proves from M1-3, 6-10 the following equivalences:
Proofs of M1-3, 6-10:

40. L[J(p⊃q)≡J(p⊃q)	L(p≡p)
41. L[J(p⊃q)≡(Jp⊃Jq)]	M8, 40, gives J7
42. L[J(p⊃0)≡(Jp⊃J0)]	41, q/0
43. L(J0≡0)	L(J0≡J0), M7: Jp⊃p, p/0
44. L[J(p⊃0)≡(Jp⊃0)]	42, 43; if Jp⊃J0 and J0≡0, then Jp⊃0 in consequent
45. L(J~p≡~Jp)	44, Df. ~ = J5, J6
46. L[(J0⊃J0)≡(0⊃0)]	L(p⊃p)≡(q⊃q)
47. L[J(0⊃0)≡(0⊃0)]	41, 46
48. L(JLp≡Lp)	M3, 43, 47, gives J3: L(Lp⊃Jp)
49. L(LJp≡Jp)	M10; L(L0≡0) L[L(0⊃0)≡(0⊃0)] de M1-3, gives J4.

Finally, Prior gives the proof from RL, L1-3, A4, 30, Df. T

50. L(n⊃p)⊃ L[a⊃L(n⊃p)]	Lp⊃L(q⊃Lp) en S5
51. L[(a⊃L(n⊃p)⊃[Ma⊃ML(n⊃p)]	L(p⊃q)⊃(Mp⊃Mq) en S5; we introduce M in consequent
52. L[a⊃L(n⊃p)]⊃ML(n⊃p)	51, A4: Mα
53. L[a⊃L(n⊃p)⊃L(n⊃p)]	52, MLp⊃p; MLp = Lp
54. L[(a⊃L(n⊃p)≡L(n⊃p)]	50, 53; p.c.
*55. TaJp≡Tnp	54, 30, Df. T = UT5[46].

The Prior´s J is: "the passing present", "the moving present", "the traveling now"[47]. The Findlay´s axiom: JxVPxVFx⊃FPx, all events, past, present,

[45] See *PTT* (2003: 192).

[46] UT5 of Kamp is modified with a constant *n* for a particular instant. Prior introduces for the same UT6: TaLp≡∀bTbp.

[47] Respectively W. R. Dennes, in California, University, Philosophical Union, *The Problem of Time*, Berkeley, Calif., (1935: 103); I. Stearns, in *Review of Metaphysics,* 4

and future, *will* be past. "We may speak as if the perceiving mind were stationary while time flows by like river, with the flotsam of events upon it"[48]. The time´s quality is to be or not to be as the "now". We live at the present and do not live simultaneously. But with Meredith the world is not necessary. "Certainly, to be the case is to be the case in the actual world"[49] and then

There are really such objects as possible worlds, and what we loosely describe as propositions of modal logic are in fact predicates of which these objects are subjects. (…) The world "actual" must be regarded as having the concealed egocentricity which the temporal "present" is sometimes said to have; the "actual" world is just the world in which *we* figure. (…) We all as it were perform on several stages at once, and in each world, we are largely ignorant of our performances -real though they be- in others. (…) So, the "actual" world is not strictly identifiable as "our" world -there are other worlds that are too (though perhaps not *all* worlds are that). The "actual" world is just the one we indicate by waving our arms about in a vague way and calling it "this" world. Alternatively, we could follow a suggestion of David Lewis and say thar each of us figures in one world only but may have a "counterpart" or counterparts in other worlds, and that our knowledge of what we might have been doing is an awareness in one world of what our counterpart is doing in another.
In either of its forms, this seems a tall story…[50]

(1950: 103); Santayana: *Realms of Being,* in *Works,* Vol. 14, p. 254; Donald C. Williams, "The Myth of Passage", p. 98.

[48] Donald C. Williams, *op. cit.,* p. 104.

[49] Prior, "World, Times and Selves", *PTT* (2003: 244).

[50] Prior, "Modal Logic ant the Logic of Applicability", *PTT* (2003: 283).

XV

London Underground

The system that fits this model is realist-modal-temporal. Is not a branching model. G.E. Hughes and M.J. Cresswell in *A New Introduction to Modal Logic*, p. 21:

"We can think of the modal game in this way: in any setting the players represent conceivable state of affairs -or, as they are often called, alternative *possible worlds*, (...) the players each player is allowed to see represent the state of affairs which are possible relative to the state of affairs which that player represents; and the letters on a player´s sheet represent the propositions that are true in that state of affairs.

Raising a hand and keeping it down represent respectively truth and falsity in the state of affairs the player represents. Hence- what K-validity of wff means is that that wff would turn out to be true in every conceivable state of affairs, no matter what propositions we were to replace its variables by, no matter what was true or false in that state of affairs, and no matter what states of affairs were possible relative to that one".

Suppose the game is to go by metro: each line represents a possible world in which each player goes. Every route is a state of affairs. And it can be applied in the same terms. The players each player is allowed to see represent the state of affairs which are possible relative to the state of affairs which that player represents; and the routes on a player´s lines represent the propositions that are true in that state of affairs. This model is realist-modal-temporal. This model fits well with Prior´s "Now". See you this model:

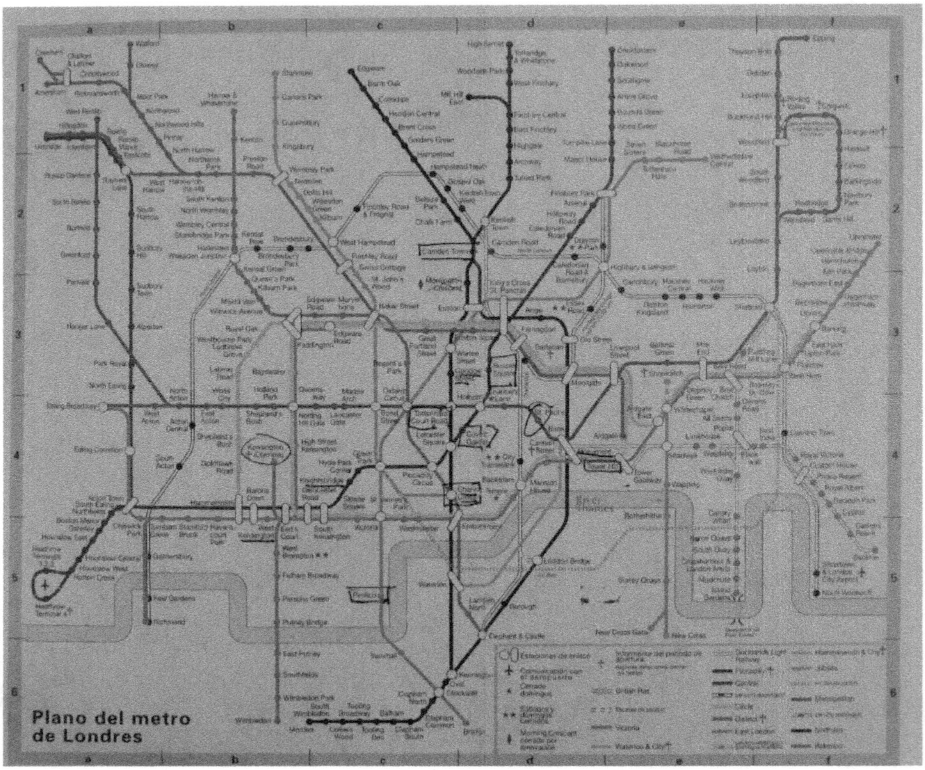

On 2 April 1801, the British squadron defeated the Danish at Copenhagen undoing Bonaparte´s alliances with the Baltic powers. Many counterfactuals arise here. But we stick with one. If Nelson had obeyed Admiral Hyde Parker´s orders, he would have been cannonaded and torn to pieces by the crossfire of the Danish guns. Nelson disobeyed Hyde Parker and not only saved his life but won the battle. It is Aristotle *De interpretatione* IX, revisited. Xerxes I vs. Themistocles. Salamina´s Battle. Suppose that Nelson and Edward Riou play chess or travel on a tube line. If Nelson goes on the violet line from Heathrow terminal to Cockfoster via King´s Cross and Riou goes on the blue line from Brixton to Walthamstow Central, he will have to go through King´s Cross. King´s Cross (Copenhagen). The result is open, but this clash is inevitable... David Lewis writes: "Our Future is the one that is part of the same world as ourselves. It alone is connected to us by the relations the (strictly or analogically) spatiotemporal relations, or perhaps natural external relations generally - that unify a world. It alone is influenced causally by what we do and how we are in the present"[51].

[51] David Lewis (1986: 208).

XVI

The City

Van Fraassen writes that the time is a logical space I add like the London Underground and city of London map. First, we could take our logical space the real straight line or a segment, or the whole of the enlarged natural numbers. In such case, we would say that the time has a beginning or that the time is topological closed[52]. Van Fraassen points out that the topology remits us to the geometry and therefore: "We may distinguish between questions of topological structure and of the metric"[53]. And Arthur Prior: "It is possible to construct a formal calculus expressing the logical features of earlier-later relation between events. But this earlier-later calculus is only convenient but indirect way of expressing truths that are not reality about events but about *things* are doing, have done, and will do"[54]. The calculation can be expressed using the structure of London underground or the map of the city. For example: Draw a dot on a blank piece of paper. Make a ball with the paper. Unfold the paper. The dot on the paper is you. The wrinkles in the paper are the relief, the geography. Draw on that relief the lines that converge on you. If you name the folds of the relief you have a map. You are that point in the present as in GPS. This is the topological structure of time. It is not a branching model but a linear one.

---

[52] Van Fraassen (1978: 129-130).
[53] Van Fraassen (1978:76).
[54] A.N. Prior, "A Statement of Temporal Realism". Edited by Jack Copeland (1996: 45, 46).

XVII

The Road Map

Arthur Prior designed a model like this in 1957:

In Appendix A [Prior 1957, pp. 119-120] Prior proposes a forked right-left logic like a road map:

"Consider, for example, a place-logic in which we have the means of formulating the law 'If it is the case (here) that *p*, then it is the case *n* miles away to the left that it is the case *n* miles away to the right that *p*'

Using the form *Lnp* for 'It is the case n miles away to the left that *p*', and *Rnp* similarly for 'right',

this law would be symbolized as $p \supset LnRnp$. An if we regarded this L-R logic as a *Tx* logic [*T* is a dyadic operator of the form 'It is (was) the case on (in, with) ___ that ___', having for its first argument some sort of name (of a time, of a place, of an individual) and for its second argument a statement, or something very like a statement] with *n* for *x* and *L* and *R* for two types of T, and knew of no way of eliminating double *Tx-ing* but that of treating all of a sequence of *Tx* prefixes but the one next to *p* as vacuous, we would conclude that place-logic is not collapsible. For the accumulation of *L-R* operators is certainly not vacuous in this logic -*LmRn*, and even *RmRn,* are certainly not equivalent to the simple *Rn*. (…) Our *x*'s are of the form λn, where this means 'the place *n* miles to the left', with *Lnp* for $T\lambda np$ with positive values of *n* and *Rnp* for $T\lambda np$ with negative values of *n*. And although we cannot equate

'Is the case *m* miles to the left that it is the case *n* miles to the left that p' ($T\lambda mT\lambda np$)

With

'It is the case *n* miles to the left that p'

($T\lambda np$), we can equate it with something of the form $T\lambda lp$, namely the $T\lambda lp$ in which *l* is the algebraic sum of *m* and *n* (with a wider range of directions, the vector sum).

As a genuine example of *Tx* logic in which this condition of collapsibility is not met we may take a tense-logic in which ln means 'the time *n* days back', *Pnp* is defined as *Uλnp* for positive and *Fnp* as *Uλnp* for negative values of *n*, and in which we assign to future contingents a third truth-value. For in such a system we cannot equate

'It was the case *m+n* days ago that it will be the case *n* days hence that *p*', i.e. *PSmnFnp*, o *Uλ(m+n)Uλ(-n)p*, with

'It was the case *m* days ago that *p*',

i.e. *Pmp*, o *Uλmp*. For even if it was the case *m* days ago that *p*, it might not have been true *m+n* days ago that it was going to be the case *n* days later that *p*, for *m+n* days ago the issue might still have been indeterminate. But if *Uλ (m+n)Uλ (-n)p* were equivalent to any singular *λp* prefix it would be clearly be to *Uλm*. So this condition of callapsibility cannot be met by this type of tense logic".

Arthur Prior worked as a radio operator from November of 1942 to January of 1945, for a Royal New Zealand Air Force radio pilot the disposition of a point on a grid of parallels and meridians was a simple matter. A logic of right and left of meridians and parallels as a Cartesian axis must have impressed the young Kripke who, after reading *Time and Modality*, raised the question of branching instead of bifurcation while working on A Completeness Theorem för Modal Logic.

Hence, the model is not Ockhamist. Suppose we drive a car to go home. What model do we have? It is not a branching model. We have a model like this:

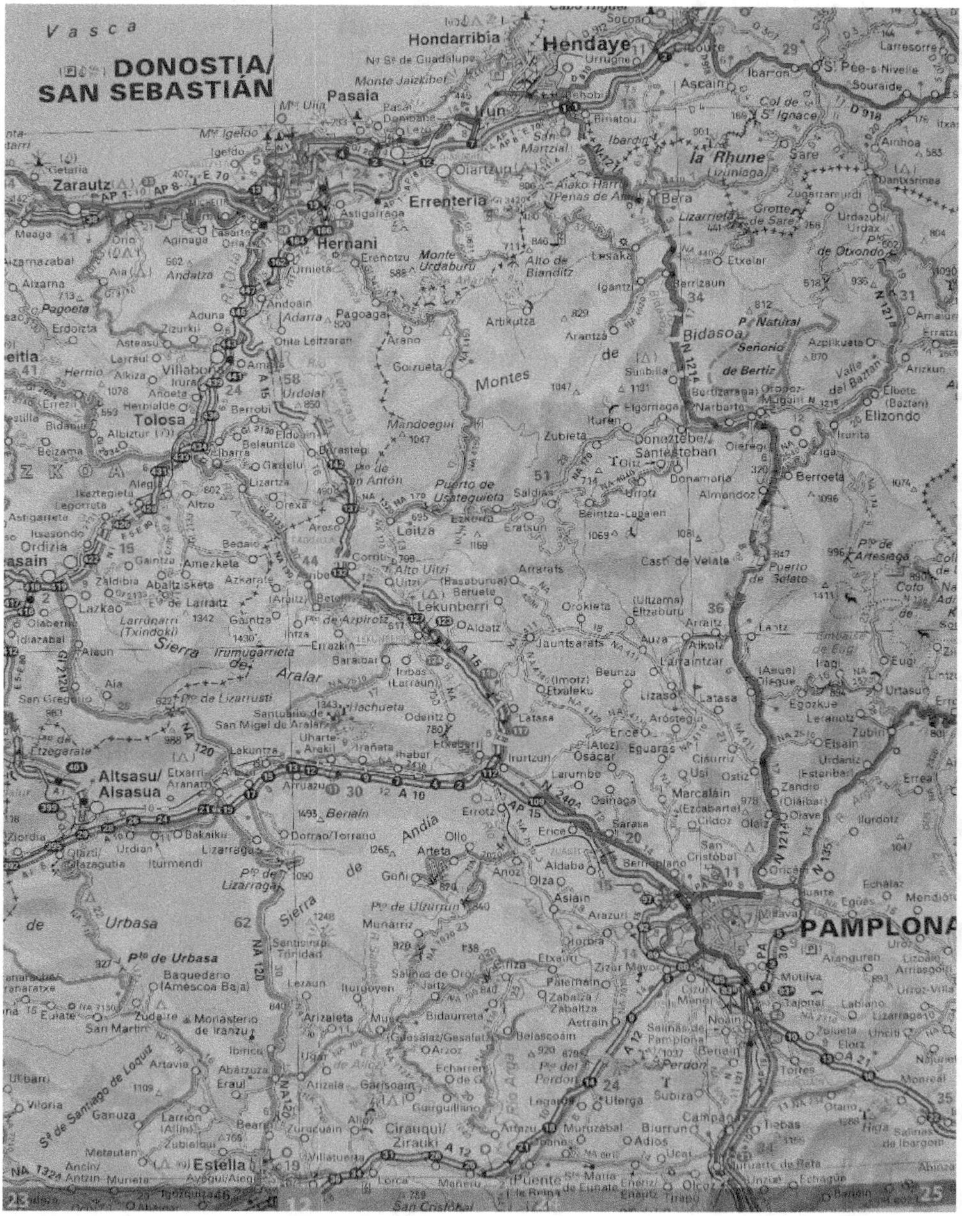

It is a linear, not a branching model. Let´s suppose we are between Pamplona and San Sebastián. Where are we? For example, in Alsasua. Our route to San Sebastián is open. We are crossed by a grid of roads and highways everywhere. When I choose a route, others do not cancel.

The future and the past do not disappear but intersect in the present. The model is indeterministic because I can choose. The model is deterministic because I cannot leave the map. It's like a chess model. What is my choice?

1. Representation of the available alternatives.
2. Evaluation of alternatives.
3. Choice of one of them.

This is not determinism. According to Storrs McCall, *A model of Universe*: "We can deliberate about is causally possible but not about what is causally necessary or causally impossible, or at least about what we know to be necessary or impossible"[55].

Again, David Lewis expresses: "If we stay with the simple account of how worlds are unified, we will conclude that where there is branching, there is one single world composed of all the branches. That would not be branching *of* worlds but branching *within* worlds; and so, the overlap of branches would not be overlap of worlds. Branching within worlds, I think, is to be accepted: it is possible that the space time of a world might have such a shape, and if it that is a possible way for a world to be then it is a way that some world is"[56].

And Arthur Prior wrote in "Modal Logic and the Logic of Applicability": "So the 'actual' world is not strictly identifiable as 'our' world -there are other worlds that are that too (though perhaps not *all* worlds are that). The 'actual' world is just the one we indicate by waving our arms about in a vague way and calling it 'this' world".

My route to San Sebastián is one world among others. The only one which I say: I arrived.

---

55 Storrs McCall (1996: 254-257).

56 David Lewis (1986: 209).

XVIII

The Folder Condition (Two)

Folder and unfolder condition are a Metaphysics concept about time and reality based in the Prior´s J and the Meredith´s δ. Let´s see:

1°) Every "now" divides the reality in worlds.
2°) The actual state of universe is multiple. There are a lot of worlds.
3°) Simultaneity existence of many worlds.
4°) There is no present.
5°) Every event is realized in one world or another.

We can define the fold condition with the theorem of Kt: $LH(p \wedge q) \equiv LHp \wedge LHq$ this theorem in K give a principle of distribution de LH in conjunction. Proof:

$LH(p \wedge q) \equiv LHp \wedge LHq$

(1) $(p \wedge q) \supset p$	p.c, 1
(2) $H(p \wedge q) \supset Hp$	Derived Rule 1, 1 $\vdash A \supset B, \vdash HA \supset HB$
(3) $LH(p \wedge q) \supset LHp$	Derived Rule 2, 2 $\vdash A \supset B, \vdash LA \supset LB$
(4) $(p \wedge q) \supset q$	p.c. 4
(5) $H(p \wedge q) \supset Hq$	DR1, 4
(6) $LH(p \wedge q) \supset LHq$	DR2, 5
(7) $LH(p \wedge q) \supset LHp \wedge LHq$	p.c. 3, 6 $A \supset B, A \supset C \vdash A \supset (B \wedge C)$
(8) $p \supset [q \supset (p \wedge q)]$	p.c. 8
(9) $Hp \supset H[q \supset (p \wedge q)]$	DR1, 8
(10) $LHp \supset LH[q \supset (p \wedge q)]$	DR2, 9
(11) $H[q \supset (p \wedge q)] \supset [Hq \supset H(p \wedge q)]$	A3. $H(A \supset B) \supset (HA \supset HB)$, p/q, $q/p \wedge q$, 11
(12) $LH[q \supset (p \wedge q)] \supset L[Hq \supset H(p \wedge q)]$	DR2, 11
(13) $LH[q \supset (p \wedge q)] \supset [LHq \supset LH(p \wedge q)]$	L1 12, p/Hq and $q/H(p \wedge q)$
(14) $LHp \supset [LHq \supset LH(p \wedge q)]$	p.c. 10, 13
(15) $LHp \wedge LHq \supset LH(p \wedge q)$	p.c. 14 $p \supset (q \supset r) \vdash (p \wedge q) \supset r$, Import.
(16) $LH(p \wedge q) \equiv LHp \wedge LHq$	$\equiv$ *def.* 7, 15.

Q.E.D

And for G, $LG(p \wedge q) \equiv LGp \wedge LGq$. Also, we have Kt: $(Hp \wedge Pq) \supset P(p \wedge q)$, it reinforces the concept of Pastness simultaneity, $Hp \wedge Pq \supset P(p \wedge q)$.

Proof:

Kt: $Hp\wedge Pq\supset P(p\wedge q)$

(1) $H(p\supset\sim q)\supset(Hp\supset H\sim q)$	Kt $H(p\supset q)\supset(Hp\supset Hq)$, p/p and $q/\sim q$
(2) $Hp\supset[(H(p\supset\sim q)\supset H\sim q)]$	p.c. Premise Mutation
(3) $Hp\supset[(\sim H\sim q\supset\sim H(p\supset\sim q)]$	p.c. 2, Transp.
(4) $Hp\supset[(Pq\supset\sim H(p\supset\sim q)]$	$\sim H\sim = P$, 3
(5) $Hp\supset[(Pq\supset P\sim(p\supset\sim q)]$	$\sim H = P\sim$, 4
(6) $Hp\supset[(Pq\supset P(p\wedge q)]$	p.c., 5
(7) $Hp\wedge Pq\supset P(p\wedge q)]$	p.c., 6

Q.E.D

and the same for $Gp\wedge Fq\supset F(p\wedge q)$.

The Theorems hold the concept of simultaneity, at the same time in Past $P(p\wedge q)$ and Future $F(p\wedge q)$. The fold condition is comparable with the notion of simultaneity. It is the concept of worlds simultaneity and times.

XIX

The Deviation of Now

Nelson´s dilemma was to obey or not to obey. Let´s formalize: O (obey), V (victory), D (death). The truth of $O \supset \sim V$ y $O \supset (\sim V \wedge D)$ y $\sim O \supset (V \wedge \sim D)$ consists in the fact that each *W* (World) in the model is a $\sim V$-*W*1, or $\sim V \wedge D$-*W*2, or $V \wedge \sim D$-*W*3. Since $\sim O \supset V \wedge \sim D$ is true in the present world, the others are counterfactuals belong to the past of the model dating from 2 April 1801. We see how the past is considered in the evaluation of counterfactuals. Returning to Nelson´s example, *W*1 and *W*2 in which Nelson obeys Hyde Parker were part April 2, 1801, and they are possible worlds. It may be that the SE wind was not blowing that morning, or that at midday in the heat of the naval battle it began to blow from the north. In others, Edward Riou who at 11,30 approached Trekroner´s fortress, would have saved his like, if like Nelson disobeying Hyde Parker, he had not headed north and suffered the fire of Trekroner that split him in two. Or that once the cannonaded ceased at 14,30 Crown Prince Frederick would not have surrender to Nelson´s threat to destroy the city. Or in others, Nelson´s echelon tactics would not having successful, if the English ship had not interfered with each other in the entrance to the King´s Passage. All the worlds in which Nelson could have lost the battle are entirely possible. The truth or falsity of *W*1, *W*2, and *W*3 depend on the deviation of "now".

"Now" is a ticket inferential. Each movement is registered on the world line like an index of instant. Each world is in each now:

A world-state proposition in the tense-logical is simply and index of an instant, I would like to say that it is an instant, in the only sense in which "instant" are not highly fictions entities. To be the case at such-and-such is simply to be the case in such-and-such a world; and that in turn is simply to be the case when such-and-such a world proposition is the case. In this sense of "instant" it is a tautology that a world proposition is true at one instant only (it is true only when *that* world-proposition is true) and so is as "strong" as any proposition that is ever true can be"[57].

We can formalize it as follows: If *D* (W3), and *J* (now):

57 Arthur Prior (1967: 188, 189), and *PTT*, XI "Tense Logic and The Logic of Earlier and Later" (2003: 117-134). Also see Peter Øhrstrøm and Per F.V Hasle "Tense Logic and Counterfactual Reasoning" (1995: 219-229) and finally Rescher and Urquhart (1971: 179).

$(\exists J)[(DJ\wedge(\forall J')(DJ'\supset J=J')] \dashv\vdash (\exists J)DJ\wedge(\forall J)(\forall J')(DJ\wedge DJ'\supset J=J')$

We will do it from left to right; we will use the rules of quantification theory:

1	(1) $(\exists J)[(DJ\wedge(\forall J')(DJ'\supset J=J')]$	Assumption
2	(2) $Da\wedge(\forall J')(DJ'\supset a=J')$	typical case
2	(3) Da	E∧ 2
2	(4) $(\forall J')(DJ'\supset a=J')$	E∧ 2
2	(5) $Db\supset a=b$	E∀ 4
2	(6) $Dc\supset a=c$	E∀ 4
2	(7) $Db\wedge Dc\supset b=c$	c. p. 5, 6; E= 5, 6.
2	(8) $(\forall J')(Db\wedge DJ'\supset b=J')$	I∀ 7
2	(9) $(\forall J)(\forall J')(DJ\wedge DJ'\supset J=J')$	I∀ 8
2	(10) $(\exists J)DJ$	I∃ 3
2	(11) $(\exists J)DJ\wedge(\forall J)(\forall J')(DJ\wedge DJ'\supset J=J')$	I∧ 9, 10
1	(12) $(\exists J)DJ\wedge(\forall J)(\forall J')(DJ\wedge DJ'\supset J=J')$	E∃ 1, 2, 11.

The following from right to left:

1	(1) $(\exists J)DJ\wedge(\forall J)(\forall J')(DJ\wedge DJ'\supset J=J')$	Assumption
1	(2) $(\exists J)DJ$	E∧ 1
1	(3) $(\forall J)(\forall J')(DJ\wedge DJ'\supset J=J')$	E∧ 1
4	(4) Db	Assumption
5	(5) Da	typical case
1	(6) $(\forall J')(Da\wedge DJ'\supset a=J')$	E∀ 3
1	(7) $Da\wedge Db\supset a=b$	E∀ 3
4,5	(8) $Da\wedge Db$	I∧ 4, 5
1,4,5	(9) $a=b$	MP 7,8
1,5	(10) $Db\supset a=b$	Condition Proof 4,9
1,5	(11) $(\forall J')(DJ'\supset a=J')$	I∀ 10
1, 5	(12) $Da\wedge(\forall J')(DJ'\supset a=J')$	I∧ 5, 11
1, 5	(13) $(\exists J)[(DJ\wedge(\forall J')(DJ'\supset J=J')]$	I∃ 12
1	(14) $(\exists J)[(DJ\wedge(\forall J')(DJ'\supset J=J')]$	E∃ 2,5, 13

Q.E.D

Each moment is unique. Nelson´s Decision to disobey Hyde Parker, the drive to San Sebastian or the naval battle at King´s Cross: that is the deviation of "now" in W3.

If we fold the universes like a sheet of newspaper, we will have the exact configuration of the universe-worlds in relation to each temporal instant in a single set of time. Reality would have a folded form of universes and times like an accordion.

XX

The Summa Ekliptika[58]

Professor David Favrholdt puts it correctly about relationship between Ernst Mach and Niels Bohr: "Mach was a phenomenalist, Bohr was not. Mach denied the existence of atoms. No micro-world existed, only sense-impressions ('Element'). Bohr, on the contrary, was a realist. I know that some doubt this and often refer to the following statement: When asked whether the algorithm of quantum mechanics could be considered as somehow mirroring an underlying quantum world, Bohr would answer: There is no quantum world. There is only an abstract quantum physical description. It is wrong to think that the task of physics is to find out how nature *is*. Physics concerns what we can say about nature'. This is often interpreted to mean that Bohr was an instrumentalist and anti-realist (and to this end, the question put to Bohr in the quotation is very often left out) but what he is saying here is that there is no *ready-made* quantum-world which is *mirrored* by quantum mechanics. Bohr always maintained that the atomic world was there, whether 'observed' or not"[59].

Physical reality is a quantum state of indeterminacy. What state is a coin in before it comes up heads or tails? It is a heads/tails state. When we roll a pair of dice, what happens the numbers that don´t come up? What state is the dice in before the outcome? The dice are multiple, that is, 36 states of 6. Arthur Prior quotes Charles Peirce: "When we say that all possible throws of a pair of dice one thirty-sixth will show sixes, the collection of possible throws which have not been made is a collection of which the individual units have no distinct identity. It is impossible so to designate a single one of those possible throws that have not been thrown that the designation shall be applicable to only definite possible throw; and this impossibility does not spring from any incapacity of ours, but from the fact that in their own nature those throws are not individually distinct. The possible is necessarily general and no amount of general specification can reduce a general class of possibilities to an individual case. It is only

[58] In about 1905 Edgar Rubin organized a group of students in a club called "Ekliptika" because the initial number of members was twelve. Among the participants were the Bohr brothers, the Nørlund brothers, Edgar Rubin, Brønland and others. "The Ekliptika Question" is the question by the Bohr-Heisenberg principle.

[59] David Favrholdt, *Niels Bohr* (1992: 123).

actuality the force of existence, which burst the fluidity of the general and produces a discrete unit…[60]"

We have one between thirty-six possible worlds, but only one outcome. It is like the movement of the "knight" on the chess board. It advances in "L" one square on the column or line on which it stands and then two squares on the perpendicular line or column. From d4 square it can occupy eight squares. The further to the edge of the board, the more this number decreases. From an angled square, it has only two moves left. On each move, unlike the bishop, the color of the square the square changes. Well, the eight moves are eight possible worlds, as in the roll of the dice each is virtual but only one is the result. What quantum state is the knight in before the movement?

An indeterminate truth-value which may be called *indeterminate* can be introduced. "The probability character of quantum mechanical predictions entails an absolutism of the individual case; it makes the individual occurrence unrepeatable, irretrievable. We express this fact by regarding the unobserved value as indeterminate, this word being taken in the sense of a third truth-value"[61].

The quantum state is a negative state of reality like the film of a camera. ½ is not the indeterminate state of trivalent logic. In fact, Łukasiewicz's system does not work as in quantum logic. ½ it works more line an understood "anti-concept" as Martin Prior, i.e. inversion. "The quantum mechanical truth-value indeterminate, however, represents a topologically different category"[62]. Hans Reichenbach quotes Paulette Février who understands the third truth-value not considered as indeterminacy, but as, "reinforced falsehood" or absurdity. "She makes a distinction between 'propositions conjuguées" and 'propositions non conjuguées'".

In 1670-71, Jan Vermeer painted the Lacemaker. A young woman embroiders a piece of lace. Her fingers cast precises stitches, offering us a reticular model like that quantum physics. Neither a straight line or a circle, but a network of lace, a convergence of multiple coordinates intertwined quantumly. On the table, pieces of red and white threads. Their seam is their planetary miniature. Our model is reticular. An "event" is something thar precedes, follows, or overlaps with something. Each stitch of "the lace" is an "event", we will also call it a "possible world". No stitch is neutral, that is, ½. Its seam is a chessboard. The move is on the player, not

[60] Prior (1957: 114).
[61] Hans Reichenbach (1944: 145).
[62] Hans Reichenbach (1944: 148).

on the board. If, as the Bohr-Heisenberg principle points out, there is a quantum ghost state, we can only describe half of reality. The hidden face of the moon remains undisclosed. Our representation is in negative, an X-ray photograph, multifocal and cubist.
It was not until 1926 that Heisenberg formulated his famous indeterminacy principle about the impossibility of measuring the position and velocity of a particle. We can formulate it with ½ (indeterminacy) as follows:

"Is double necessary that [If A is possible and B ½, entails that A=B], and we have A possible, and A ≠B, then we don´t have ½B".

We have quantic entanglement: $LL(MA \wedge 1/2B \supset A=B)$, MA, $A \neq B \vdash \sim 1/2B$

Proof:

1. $LL(MA \wedge 1/2B \supset A=B)/1$	(Premise,1=quantic state)
2. MA/1	(Premise)
3. $A \neq B/1$	(Premise)
4. $\sim\sim 1/2B/1$	(Negative conclusion)
6. 1/2B/1 (de 4)	(Double Negative)
7. B/1 (de 4)	(1/2B = B, by D.A. Bovchar[63])
8. $1 \models\Rightarrow 2$	(quantic entanglement)
9. A/2	(de 2 y 8).
10. $L(MA \wedge 1/2B \supset A=B)/2$	(de 1 y 8)
11. $2 \models\Rightarrow 2$	(Reflexive)
12. $(MA \wedge 1/2B) \supset A=B/2$	(de 10 y 11)

Now we have branching

13. $\sim(MA \wedge 1/2B)/2$	14. A=B/2
	15. $2 \models\Rightarrow 1$ Quantic entanglement
	16. A=B/1 Branch closed (3 y 16)

Now we have branching

17. ~MA/2	20. ~1/2B/2
18. L~A/2	21. ~B/2 (de 20)
19 ~A/2 de 11 y 18) Closed by 9 y 19)	22. $2 \models\Rightarrow 1$ 23. ~B/1. (Branch closed by 7 y 23)

[63] "El Sistema trivalente de Bovchar y las paradojas". Notebook unpublished. Juan Carlos León. Universidad de Murcia.

Suppose someone gives a friend a pair of gloves. A is John and B is Diane. A gives B some gloves but forgets one on the table. A is in London and B is in Reykjavik. The event is interwoven but incomplete. Until we open it, we will not know whether it is the right or the left one. However, whether it is one or the other, they will influence each other. Something similar happens with quantum entanglement.

$$LL(MA \wedge 1/2B \supset A=B),\ MA,\ A \neq B \vdash \sim 1/2B$$

Is reciprocal, nor asymmetric, accordance with Storrs McCall: "hold outside the light cone, between spacelike separated events; 3. Instantaneous and non-local; 4. Not explicable by a common cause; 5. Direction of influence dependent on choice coordinate frame. Plainly, non-local effects differ radically from ordinary causation"[64].

[64] Storrs McCall (1994: 108).

Bibliography

Barrow John D, (1994), *Teorías del todo.* Traducción de Rosa Álvarez. Título original *Theories of Everything. The Quest for Ultimate Explanation.* Oxford University Press (1991), para la edición castellana, Crítica, Grijalbo, Barcelona.

Borges, Jorge Luis (1977), *Historia de la noche*, en Obra Poética 3, Alianza Editorial.

Copeland Jack (2002), "The Genesis of possible worlds semantics", *Journal of Philosophical Logic.*

Favrholdt David (1992), *Niels Bohr Philosophical Background,* Historik-filosofiske Meddelelser 63. The Royal Danish Academy of Siciences and Letters. Commissioner: Munksgaard, Copenhagen.

Findlay J. N. (1941): "Time: A Treatment of Some Puzzles". In *The Philosophy of Time.* A Collection of Essays. Edited by Richard M. Gale. Anchor Books, New York.

Gardies, Jean- Louis (1979), *Lógica del tiempo.* Traducido por Javier Ordóñez. Revisado por Pascual Martínez Freire. Presses Universitaires de France. De la edición española, Paraninfo, Madrid.

Gordon Childe (1996), *Los orígenes de la civilización.* Breviarios, Fondo de Cultura Económica, 92.

Horwich Paul (1987), *Asymmetries in Time.* A Bradford Book, The MIT Press, cuarta impression, 1992, (Cambridge, Mass).

Lewis David (1986), *On the Plurality of Worlds*, Basil Blackwell, New York.

Łukasiewicz, Jan, "On Determinism", (1961), in Storrs McCall (1967), *Polish Logic 1920-1939.* Oxford Clarendon Press.

Meredith C. A. and Prior A. N. (April 1965), "Modal Logic with Functorial Variables and a Contingent Constant", Notre Dame Journal of Formal Logic, vol. 6, no. 2.

McCall Storrs (1996), *A Model of The Universe.* Oxford Clarendon Press.

Prior Arthur Norman (1957), *Time and Modality.* Oxford University Press.

Prior, Arthur Norman (1967) *Past, Present and Future.* Oxford University Press.

Prior, Arthur Norman, (1968) *Papers On Time and Tense,* XI "Tense Logic and The Logic of Earlier and Later". Oxford Clarendon Press.

Prior, Arthur Normam (1969). *Papers on Time and Tense.* "Limited Indeterminism". Edited by Per Hasle, Peter Øhrstrøm, Torben Braüner, and Jack Copeland. Oxford University Press, (2003).

Prior, Arthur Norman "Modal Logic ant the Logic of Applicability", *Papers on Time and Tense* (2003).

Prior Arthur Norman, "World, Times and Selves", *Papers on Time and Tense* (2003).

Prior A. N, "A Statement of Temporal Realism", in *Logic and Reality*, Essays of Legacy of Arthur Prior. Edited by Jack Copeland (1996).

Øhrstrøm Peter y Hasle Per (1995), *Temporal Logic, From Ancient Ideas to Artificial Intelligence.* Kluwer Academic Publishers, Dordrecht, Boston, London.

Rescher Nicholas and Urquhart Alasdair (1971), *Temporal Logic*, Springer-Verlag, New York, Wien.

Reichenbach Hans (1944), *Philosophical Foundations of Quantum Mechanics*. University of California Press. Berkeley and Los Angeles.

Van Fraassen C. A., (1978) *Introducción a la Filosofía del tiempo y del* espacio.

Van Fraassen, Bas C. (1980) *The Scientific Image*. Oxford University Press.

Williams, Donald C. "The Myth of Passage". In *The philosophy of Time. A Collection of Essays.* Edited by M. Gale, Anchor Books, New York, 1967

Wittgenstein Ludwig (1921), *Tractatus Logico-Philosophicus*. Translated by D. F. Pears and B. F. McGuinness. Routledge & Kegan Paul, 1961, 1974. First published in Routledge Classic, 2001.

INDEX

www.ingramcontent.com/pod-product-compliance
Lightning Source LLC
LaVergne TN
LVHW022321200926
848481LV00004B/1393